AFTERSHOCK

The Loma Prieta Earthquake and its Impact on San Benito County

By
James Z. McCann

Photo Editor
Ray Pierce

Seismic Publications

Hollister, California

Aftershock: The Loma Prieta Earthquake
and its Impact on San Benito County

Written by
James Z. McCann

Photo Editor
Ray Pierce

Text Editors
James Z. McCann
Adele Churchill

Technical Adviser
Jeff Tolhurst

Production Manager
Jim Churchill

ISBN 0-9628177-0-8
Library of Congress
Cataloging Number 90-63594

First Printing, November 1990

Printed in the USA

*Cover photo by Eric Raper. Photo
on page 4 by Ray Pierce, Free Lance.
Photo on page 80 by Craig Cochran,
San Benito Sun.*

**Seismic
Publications**

**642 San Benito St.,
Hollister, CA 95023 USA**

This book is dedicated to the thousands who volunteered their time, energy or financial support to the earthquake relief and recovery efforts in Hollister and San Benito County

ACKNOWLEDGMENTS

A book such as this could only have been made with the cooperation and support of hundreds of people. The response to our public call for earthquake photos and stories was almost overwhelming.

The process of selection, both of the photographs and commentary, was a difficult one. Many photographers turned in pictures of the same scenes, for example, and only one was chosen. Likewise, most everybody's experiences during and after the quake were interesting or touching and sometimes amusing, but space limitations and the desire to let the photographs tell most of the story prevented the use of all the material available.

If your photo or comments did not make it into the book, chances are your friend's or neighbor's did. The book, however, could not have been created without each of you. (A list of contributors is included on page 79.)

A special thanks goes to Suzie Crump for her organization, persistence, copy editing and smiles; and to John Haydon for his expertise, enthusiasm and general support.

This book was made possible by sponsorship from
The Granite Rock Company

Other assistance was provided by Bill Clyde and ERA Country Realty for office space and technical support, the Free Lance Newspaper and the San Benito Sun for photographs.

Thank you.

Jim and Adele Churchill
October, 1990

Contents

ABOVE: Stunned by the ferocity of the 7.1 magnitude Loma Prieta earthquake, people stare at the devastated I.O.O.F. building on Fourth Street minutes after the facade collapsed.
Photo by Ray Pierce, Free Lance

Life in Earthquake Country

By Jim McCann

It's just another earthquake."

That was the unimpressed reaction of most local people when the first jolt struck at 5:04 p.m. on a warm Tuesday evening in October 1989. The World Series pre-game show was tuned in on many television sets, dinner was in the oven for some, and thousands others were making their way home from work.

Nothing was out of the ordinary.

But the first jolt was followed by another and then another, and then the ground really started jumping. By this time those who could move were under doorways, under desks, hanging on to anything which might provide a sense of security. Most drivers thought a tire had blown. Pedestrians watched telephone poles whip from side to side, and power lines swing like jump ropes. Fifteen seconds later 170 homes in Hollister were too damaged to live in. More than 20 buildings in downtown Hollister were either destroyed or so severely hit by the earthquake they would require major repair. Power was out, and the danger of gas leaks was everywhere.

It was not "just another earthquake."

Even the immediate impacts of the Loma Prieta quake took days to realize. On a basic level, it was a *natural* disaster. The forces of nature, locked in an underground arm wrestle, briefly gave way and unleashed a tremendous amount of energy. Up on the surface, however, the result seemed entirely *unnatural*. As the nation learned of the collapse of the Cypress structure in Oakland, the Marina District fire in San Francisco, the destroyed Pacific Garden Mall in Santa Cruz, the quake's deadly effects assumed staggering proportions. In San Benito County houses crumpled as if made of toothpicks, brick walls tumbled like children's blocks, and whole businesses were destroyed in less time than it takes to ring up a sale. The seismic event was over in seconds, but the effects will be felt for years.

To the dismay of the San Benito County Chamber of Commerce, Hollister is known as "The Earthquake Capital of the World." The geology of the area assures that Hollister will always have plenty of earthquakes. It should be comforting to know the myriad small temblors felt in the Hollister area reduce the likelihood of a "major" earthquake centered here — the proverbial "Big One." It should be, but to many it is small consolation for living through something like the Loma Prieta quake. It was "big" enough.

The Quake of '89 and its aftermath quickly created a new chapter in the history of San Benito County, a story of great loss, hardship and fear. But it is also the story of courage and selfless support for friends and strangers alike — from volunteers who staffed the relief shelters to the untold thousands of people across the country who sent food, clothes and money. Their good will helped ease the pain and made the victims' recoveries a little smoother. Furthermore, the generosity of volunteers and service organizations, and the long hours of hard work by many "just doing their job," will long be remembered.

Aftershock tells the story from the perspective of those who experienced it first hand, from the newspaper photographers to the neighbors down the street. Unlike any other book published following the Loma Prieta earthquake, this volume includes many pictures taken by local residents. Those, along with the stories and comments from numerous interviews, make *Aftershock* a book by and about the people of San Benito County. This is the community's record of what happened.

To the people of San Benito County: this is your book, a chronicle of the Loma Prieta earthquake and its impact. To those readers who don't live at the intersection of three major faults, as Hollister residents do, this is a book about a community's experience with one severe example of the geologically inevitable.

Earthquakes are a fact of life every San Benito County resident must deal with. There are constant reminders, small rumbles at any time of day or night which pass quickly. Even the most seismically well-adjusted, however, were thrown for a loop by the intensity of the Loma Prieta Earthquake. That and the realization October 17, 1989 was not "The Big One."

A sunny afternoon

Some call it "earthquake weather," although it is hard to define - when the sun shines a certain way, the wind blows gently from a certain direction, and you think to yourself, "It's been awhile...." It is a feeling, a sense of something in the air. Looking back on it later, a few people recalled having such a feeling on the afternoon of October 17, 1989.

Jim Pacheco was caught in downtown traffic during the earthquake

"I was facing east in the Fourth Street intersection and I looked over to Rovella's Gym and the Odd Fellows Hall and I watched that brick building tumble to the ground, right there, with nowhere to go. A truck was blocking me in the lane, it was typical 5 o'clock traffic on Fourth Street, locked up tighter than a drum. I thought, 'Where do I go? What do I do at this point? I'm trapped.' From the angle I was looking at the building, I thought, 'It's going to hit me.' But I was very fortunate. After all the rumbling and crumbling, I came out of it with just a little bit of sheetrock dust and a couple of bricks in the back of my pickup.

"The Blazer that was parked in front of Rovella's Gym was crushed.

"I got out of the pickup, after everything stopped and the dust settled, and looked around. There were horns blowing and people thought possibly somebody was pinned in that vehicle. But, I think at that time I started to get a little spooked and a little bit nervous. Looking down the intersection and being a little claustrophobic anyway, I thought, 'What's happening here?' The wires were touching and arcing and Fourth Street was a mess.

"The only time I ever felt as afraid in my life was when I was at Watts in the riot and I was shot at. Those two occurrences were the worst in my life. I put the two of them together. It's just a 'where do you go?', 'what do you do?' kind of thing. No exit."

Leslie Ostoja, age 9

"I was sitting on the floor watching T.V., and then the T.V. got black and white and I got scared."

RIGHT: The scene on Fourth Street, minutes after the earthquake struck.
Photo by Don Smith, Free Lance

Roberta Fletes *was downtown for a meeting*
  "We saw and heard the I.O.O.F. and Human
Services building collapsing. I was sure that
the world as I knew it was ending."

Kathy Fehlman was inside the I.O.O.F. building when the facade collapsed

"I was in Rovella's Gym in the tanning booth when the big quake hit. I was laying there, the lights were all on and all of a sudden the tanning booth starts to shake. Everything started shaking, the lights went out and I realized, 'My God, this is an earthquake, and it's a bad one.' At first, as I was laying there I thought, 'well, I guess this is okay, I'm all right, and then it dawned on me, 'Wait a minute, I'm surrounded by all these glass tubes. I'd better get out of here.'

"So I lifted up the lid and felt my way out. The whole room was pitch black, dark, shaking and there was this roar going on all around me. I kept hearing bar bells and things falling. I was groping around trying to find my clothes, and it kept roaring and shaking. I'm looking and looking and trying to find my bra. I can't find it anywhere, and all of a sudden it dawns on me, 'Kathy, what does it matter about a bra? Get the rest of your clothes.' I got the rest of my clothes on, felt for my bag, and started to feel my way out of the building.

"I didn't realize the seriousness of that earthquake until I got out to the front and all in front of me there were bricks, smashed on tops of cars. As terrible as it seems, my being so vain that I wouldn't dare leave without having my clothes, that's probably what saved me, because the time it took me to find them and put them on kept me from panicking and running outside."

ABOVE: Weightlifters inside Rovella's Gym had to climb over the sidewalk rubble to escape to the street.
Photo by Craig Cochran, San Benito Sun

David Edge, County Administrative Officer and Director of Emergency Services, was in his second-floor downtown office, located in the old Wapple house on the corner of Fifth and West streets

"I was sitting at my desk, the secretaries were closing up shop. It started to roll.

"I break the experience down into thirds. The first third, for Californians anyway, was kind of, 'Okay, it's another earthquake. Where's this one going to go?

"The second was, 'This is a pretty big one.' It's the duck and cover stage.

"The third stage, I was under the desk. I had come to the conclusion the house was going to collapse."

Kathy Ritter, *Hollister, has relatives in Southern California*
"My brother heard that Hollister was demolished and over 200 people were killed. He was really concerned."

RIGHT: Everyone agrees Hollister was fortunate to have no quake-caused deaths.
Photo by Dan Craig

DO NOT ENTER
UNSAFE TO OCCUPY

BY ORDER OF THE HOLLISTER
CITY BUILDING DEPT. 10-89 -18-
VIOLATORS WILL BE PROSECUTED

ABOVE: Realizing the extent of the damage, Hollister's Building Department immediately printed the signs it posted to keep people out of dangerous buildings.
Photo by Jim McCann, San Benito Sun

Mary Schneider, *Monterey Street*
"When the quake hit, Raul Prado's construction crew was up on scaffolding on what was becoming the second floor of our house. It sounded like a train was running through as timbers creaked, tiles fell, things crashed out of cupboards, and I fought to hold on to the bathroom doorway, where the door kept swinging at me. We called to each other - 'everyone okay?' The new foundation held up, but it was more than a week before the workers felt all right about getting up on a scaffold again."

Bob Tiffany, *Tiffany Motor Company*

"I was standing in our showroom watching the World Series pre-game on T.V. with several other employees. The telecast first went out and then the quake struck a good 4-5 seconds later. I remember the delay because I had time to swear under my breath, and turn to comment that I'd have to head home to watch the game."

Bill Mifsud Jr., *co-owner of Bill's Bullpen baseball card store*

"We had just come off a very busy week at the store because the World Series was Christmas for us, both teams in the World Series. At 5 o'clock I had just turned on the T.V. set to watch the game.

"The guy on the T.V. said, 'We're having an earthquake,' and I said, 'Yeah, we sure are, because I can feel it too.' It started out relatively mild and then the whole thing just shook and things were flying. It was kind of chaotic because there was a lot of kids in the store and they were kind of screaming. I just told them to 'get on your bikes because your moms are going to try and find you.'"

ABOVE: John Borges Photography lost its awning and decorative brick facing, but the building remained strong. Borges moved his business into his home.
Photo by Dennis Taylor, Free Lance

ABOVE: The quake brought the kind of damage not often experienced in any town, large or small.
Photo by Richard Pereira

Michael Clark, *Paicines*

"I'd been having some sort of car troubles, and thought something was happening with my car. I was stopped in the intersection there by the Quick-Stop Market (on San Benito Street), and the wiggling kept going. I kept pushing harder and harder on the break pedal and the wiggling kept wiggling. There was a Cadillac next to me that was shaking enough so that I was concerned about it possibly scooting over and hitting my car.

"After everything settled down, I thought, 'That was a good one.'"

RIGHT: When the shaking stopped, employees at Mauro's Stationers on San Benito Street left the building not through the front door, but through the shattered windows.
Photo by Jim McCann, San Benito Sun

BELOW: This house on Fourth and Line streets buckled and slid off its foundation. It could not be renovated and was torn down.
Photo by Fred Lanini

Mike Mifsud, *Hollister*

"I was in Watsonville driving back to Hollister when the earthquake struck. The first thing I saw when I got into Hollister was people standing outside in front of Hollister Super Market and the hardware store. When I got to the intersection [Fourth and Line] I saw a house that was collapsed. I knew something serious had happened then.

"When I got home I found my wife, mother-in-law, one of my sons and my little girl outside, crying, in front of the house. I found a bookcase that had fallen over and almost crushed my little girl, and our son Kevin grabbed her and pulled her out of the way in time."

Gene Kohagen, *Ridgemark resident*

"It was a lovely warm, sunny day and I was sitting reading on our patio. Abbie, our mostly black lab started getting antsy and kept trying to get in my lap. The other dog, Bo, was tossing his tennis ball around. Then 'it' started and I jumped from my chair trying to get away from the sliding windows. I couldn't stand up and kept staggering, protecting myself from the cement patio with my hands. I remember Bo, who loves to play ball, trying to chase his tennis balls that were going every which way but not being able to stand and catch them.

"One thought that went through my mind was, 'Is this when we break off and sink into the ocean?' At that time I was certain it was possible."

ABOVE: Firefighters search the San Benito County Mental Health building on Fourth Street. Left to right: Joe Martinez, Mike Jackson, and Joseph Martinez. Masonry from the I.O.O.F. collapse damaged the roof and emergency sprinklers doused the interior, soaking valuable records.
Photo by Ray Pierce, Free Lance

Marilyn N. Hull, resident of Mission Oaks Mobile Home Park
   "Our clock didn't really stop at 5:04 that day for it fairly flew, and we never did find one of the hands."

ABOVE: Collapsed porches were some of the more visible structural damage caused by the earthquake. *Photo by Dennis Taylor, Free Lance*

LEFT: More than 100 mobile homes were tossed from their supports at the Mission Oaks Mobile Home Park in west Hollister. Residents there, many of them senior citizens, took care of each other, as it took two full days for word to reach local officials of the park's extensive damage. *Photo by Dennis Taylor, Free Lance*

ABOVE: The force of ground movement buckled this section of downtown Hollister street curb.
Photo by Laurie MacFarlane

BELOW: Houses in the Powell Street area suffered the most extensive structural damage.
Photo by Dennis Taylor, Free Lance

Allison Mattox, *16*

"I was sitting there, waiting for someone in my family to show up. I didn't know where anyone was.

"It was scary not knowing. There was not a sound around. The neighbors were out checking that everything was okay, and I was helping check the neighborhood and stuff. You didn't know what was going on because it was so quiet. Not a sound around."

LEFT: The Land House, on Powell Street, was the oldest home in Hollister prior to the Loma Prieta earthquake. Located in an area that may have once been a river bed, the old structure could not withstand the shaking. It has since been razed.
Photo by Eric Raper

Ralyn Monique Johnson, *13, then a student at Rancho San Justo School*

"We just got through with the volleyball game against San Juan Bautista. We won. It felt like a rollercoaster, and I was holding on to a pole and I was sitting down on the ground. Half the cheerleaders were crying and my friends screamed their heads off."

Robert Salcedo, *17*

"I had a broken knee at the time. The earthquake hit, and I tried to hop on one leg underneath the door, and my little brothers came right down the hallway, dove underneath the table, and we rode the earthquake out.

"I looked outside from around the back, into the front yard. All the telephone wires were down, the street was buckled, and there was water shooting from the middle of the road.

"At least we got three days off of school."

18

Jennifer Lierly, *whose family's Seventh Street home was destroyed by the earthquake*

"We were sitting there, my brother, a friend of ours and I. For some reason I got up and started walking across the room. The quake hit and so I went and stood in a doorway. Dave told me I was being a big baby and so then all of a sudden it really hit hard, and they jumped up and ran over there with me.

"When it was over, we couldn't get out because the doors wouldn't open."

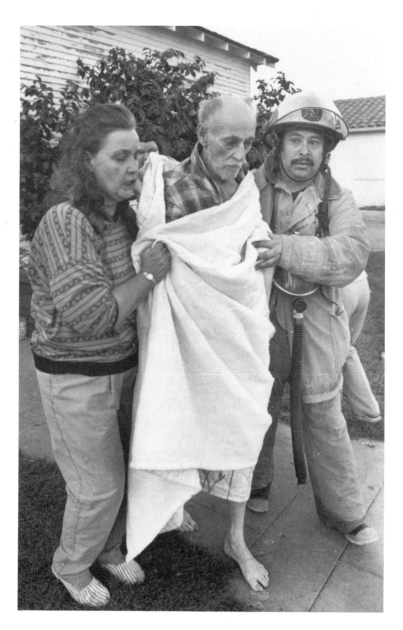

RIGHT: This gentleman was trapped in his bedroom for a short time when his home shifted on its foundation. Emergency crews responded to more than 250 calls in the four days following the earthquake.
Photo by Dennis Taylor, Free Lance

Velma Growney, *North County*

"I was cooking and I heard Joe call from the front room, 'Babe, there is an earthquake coming.' Then a big rumble, my pots and pans came tumbling off the stove. I tried to step aside so I wouldn't get burnt. The cupboard doors came flying open and all my dishes came falling out. Pictures came off the walls. The refrigerator door came open and everything fell out. Joe kept calling to me to come into the front room. I finally made it over to the doorway and found Joe on his hands and knees. He had been thrown from the recliner and was trying to get to his feet. Everything was falling, crashing and making so much noise."

Gregory M. Camacho-Light,
Hollister City Councilman

"One might think as a city councilmember that my most vivid recollections of the earthquake would consist of high powered meetings with government officials, senators, assemblymembers, the governor, or perhaps the countless council and emergency response meetings I attended.

"Frankly, when I think back on those first few days after the quake, it is those individuals at Dunne Park living in their tents, those who lost their mobile homes at Mission Oaks, and the people being housed at San Andreas School that I remember most."

ABOVE: Rather than return home to unknown hazards, many families preferred to camp out. This family chose Dunne Park.
Photo by Free Lance

BELOW: At nightfall, people gathered to keep each other company and share their common fears.
Photo by Free Lance

ABOVE: K & S Market stayed open on generator power, supplying residents with emergency essentials.
Photo by Ray Pierce, Free Lance

BELOW: The Hazel Hawkins Hospital emergency room, packed with patients soon after the quake, quieted down as the night wore on.
Photo by Ray Pierce, Free Lance

Gene O'Neill, *whose wife, Gerri, dislocated her shoulder when the refrigerator knocked her to the floor*

"The emergency room was already crowded with people, many more frightened than hurt. The staff was cool and efficient, giving out ice packs and seeing patients in order of the seriousness of their injuries. In a short time Gerri was taken to X-ray, given an anesthetic through an IV, and put back together by Dr. Jefferey Carter, an orthopedic surgeon from Monterey who was stuck there because of the earthquake. What a delicious irony!

"We were on our way home in about 40 minutes."

Assessing the damage

Margaret Houghton, *South Street*

"My home went down during the earthquake. When I say it went down, it separated and went off the foundation. I was sitting right by the fireplace and I knew from the jerk that it was going to be bad, and I'd always been told to get away from the fireplace, and go down the hall to the old-fashioned parlor where it was stronger. That's where I went.

"It seemed like seconds and it was all over. Everything was broken. I went to go out the front door, there was no porch. I went to go into the dining room and there was a three-foot gap where it had separated. I went to the side door, couldn't get out that. I got out the window."

ABOVE: Built in 1885, the historic Hawkins House, owned by Margaret Houghton, was split in two by the quake. A potential candidate for state and federal historic designation, the house took more than a year to rebuild following the Loma Prieta earthquake. As with many damaged buildings, entry was restricted immediately following the earthquake.
Photo by Craig Cochran, San Benito Sun

RIGHT: Walls sheared
as the grand old house
was ripped apart.
*Photo by
Margaret Houghton*

LEFT: Irreplaceable heirlooms
splintered as the house tore
apart. It will take more than
$265,000 to rebuild the house.
Photo by Margaret Houghton

Dana Scattini Velho,
Hollister resident

"That day [Oct. 17] I was really feeling sorry for ourselves. The next day, after driving through town, was when I realized how lucky we were. Sure we had lost a lot, but we still had our home and each other. And besides, I got all new dishes for Christmas."

LEFT: Bruce Wilson surveys the damage of his vacuum and sewing machine repair shop on San Benito Street.
Photo by Free Lance

RIGHT: John Barrett, owner of McKinnon Lumber, loaded up with plywood to distribute to dazed downtown business owners. He later snapped this photo of his lumberyard before restacking and sorting materials.
Photo by John Barrett

BELOW: Origninally known as the Opal, then the State, the art-deco Showcase Theater received $1,000,000 in structural damage. In late October, 1990, the theater burned to the ground
Photo by Free Lance

ABOVE: Jim Gibson, owner of Hollister Super Market, opened his store without power the morning following the quake.
Photo by Ray Pierce, Free Lance

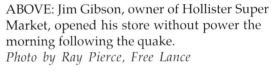

John Hardin*, Manager, Central Valley Cinemas*

"The Showcase was nowhere near as lucky as The Granada. Its damage was like none I had seen before. Not only was equipment strewn across the lobby, but huge cracks ran alongside the screen and were starting to spread across the balcony as well.

"What a sad way to lose one of Hollister's most memorable buildings."

BELOW: Javier Garcia removes inventory from his jewelry store. The building, of unreinforced masonry, was later demolished.
Photo by Ray Pierce, Free Lance

ABOVE: Anxious business owners await the opportunity to enter their cordoned-off buildings and salvage what they can.
Photo by Jim McCann, San Benito Sun

Bud Granger*, owner, Granger Printing, on going back to his San Benito Street business without permission after city officials tagged it "unsafe to occupy."*

"That Sunday when we had such a black cloud over Hollister, I mean it just looked like it was going to be a horrendous rain, I was afraid the rain was going to destroy what the earthquake hadn't already ruined. So, that's when I went up on the roof to cover the roof up. Got arrested.

"They put handcuffs on me behind my back, pushed me into the back seat of the squad car, and took me down, took my fingerprints, said they would have taken my photograph but the photo machine was broken.

"I was fit to be tied. I was so angry I could've spit on them. I thought it was so unneccessary. But, I guess they felt they had to do their job. I felt like I had to do my job."

Bill Mifsud Jr., *co-owner with his father of Bill's Bullpen, formerly on San Benito Street*

"We went back in November, I think it was about the 15th, when they were actually demolishing our building, after they deemed it irreparable. My dad took a video camera. He was videotaping the barren lot - this is where this was, this is where that was. Then he zoomed in and you could see about five or six baseball cards underneath the dirt, and he says, 'There's Rick Sutcliffe's card.'

"I remember him saying that, and you could see it in there. That was sad, but it was like the leaves falling, the change.

"It was moving."

LEFT: Damage suffered in residential areas extended beyond foundations. Landscaping was also rearranged.
Photo by Jim McCann, San Benito Sun

Loree Chappell, *Oak Street*
"A lot of friends had a lot of damage if they had one of those oval-shaped, glass-front cupboards. They all fell away from the wall and all broke all of the dishes. The doors fell open and the dishes went scooting out on the floor. Almost every one of them. Now, since that time they've bolted it or got it fastened back to the wall, which they should have had in the beginning. But they never ever thought anything would happen like that."

RIGHT: A number of sidewalks in Hollister show the dramatic effects of the Loma Prieta earthquake.
Photo by Mary Ann Meyers

Velma Newberry, *Powell Street*

"The noise of everything in the house falling was horrible. Before I could get up the living room was in shambles - floor lamps, vases, plants, figurines - all on the floor. Then the awful crash when I thought the house had split in two, but it turned out to be my fireplace chimney which had fallen."

RIGHT: Kitchen scenes like this were repeated throughout Hollister. Even this two-year old house on A Street suffered damage.
Photo by Tracy Martin

Adrian Perry Carter, *Hollister*

"I lost lots of crystal and some china, most of it 50 or more years old.

"I took a big blue plastic water bottle and filled it with the pieces and have it out on my patio - to remember the earthquake!"

Kathy Dassel, *Hollister*

"My family has always been an avid collector of antiques and collectibles, and it just makes me realize how unimportant and insignificant things like that are when earthquakes and natural disasters happen, because all I could think about was, 'I didn't care if I lost everything as long as everybody was healthy and safe. Nothing else mattered.'"

LEFT: Without lights the evening of the earthquake, many people could not clean their kitchens until the following day.
Photo by Jim McCann, San Benito Sun

LEFT: Hollister Main Street Program Director Dan Craig, left, Hollister Downtown Association President Bob Tiffany, center, and Hollister Building Director Tom Barry discuss the situation on San Benito Street. Barricading blocks of downtown, though deemed necessary by city officials until a full assessment was completed, was a source of tension. *Photo by Jim McCann, San Benito Sun*

Dan Craig, *Hollister Main Street Program Director, looking back on those tense few weeks when the fate of much of downtown was still unknown*

"It would be several months before the damage could be fully assessed. In the meantime, city officials, doing their part to protect lives and property, promptly closed off portions of downtown to traffic. Soon thereafter, actual wooden barricades were erected across San Benito Street (a state highway) to cordon off areas where buildings had been determined to be unsafe.

"The streets were not fully reopened until almost two months after the earthquake struck."

BELOW: Young California Conservation Corps workers were pressed into service throughout the earthquake-damaged areas. Here they help construct the emergency barricades used to cordon off downtown streets.
Photo by Free Lance

ABOVE: The emergency command center was immediately established on Park Hill following the earthquake.
Photo by Ray Pierce, Free Lance

BELOW: Law enforcement personnel, such as Hollister Police Officer Bob Brooks, spent many lonely hours on duty downtown.
Photo by Ray Pierce, Free Lance

Ethell Catherill, *Area Manager, Pacific Gas & Electric company*

"Employees worked around the clock for, I would say, three and four days without any sleep. There was a lot of damage to repair. We learned one thing, the first 24 hours, Hollister Utility Company, P G & E in the Hollister area, you're on your own. No communication, that's the weirdest feeling, no communications from the outside world."

ABOVE: PG & E crews were a common sight in downtown Hollister, repairing downed lines and inspecting for damaged gas pipes. As many as 150 linemen from all over the western United States came to work in San Benito County.
Photo by Ray Pierce, Free Lance

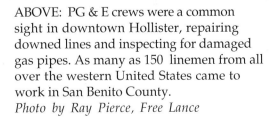

Hugh Riley, *Hollister City Manager*

"We still have a lot of work to do downtown. By in large, we were lucky, and it's presented us with some real opportunities, incentives, and I think it shook us into a reality, not only with preparedness but also with development, planning and those kinds of things."

David Edge, *County Administrative Officer and Director of Emergency Services*

"We were lucky in October (1989), and it remains to be seen how the luck will hold.

"Everybody knows there are more big quakes in the future. That's the one certainty. Our job is to make sure we have good plans in place to minimize our reliance on luck."

RIGHT: Downtown Hollister was essentially isolated following the earthquake. The barricade on the city's main street was gradually reduced from four blocks to one, as damaged buildings were demolished or repaired. The plywood around the Mode O'Day and Las Palmas buildings, foreground right, and the Showcase Theater three blocks away, was eventually replaced by chain-link fence. Those fences were still in place a year later.
Photo by Ray Pierce,
Free Lance

ABOVE: Rain dampened an already quiet downtown as the plywood barriers were erected.
Photo by Free Lance

RIGHT: The house was structurally sound, but water damage forced a temporary evacuation for this downtown Hollister family.
Photo by Free Lance

LEFT: An eight-foot plywood structure, known as "The Wall," blocked entrance into the heart of downtown Hollister.
Photo by Jim McCann, San Benito Sun

BELOW LEFT: Ice cream won't stay frozen long with the power out, as BeVo's Sandwich and Ice Cream shop discovered. BeVo's donated the thawing tubs to local relief efforts.
Photo by Linda Beavers

BELOW RIGHT: Many sidewalks popped and broke under the seismic stress. This one is on Fifth Street.
Photo by Ann Bouchard

ABOVE: K & S Market, survivor of many earthquakes, has arranged the bulk of its canned goods in chute-like fixtures. The pet food aisle was not so fortunate.
Photo by Ray Pierce, Free Lance

ABOVE: Most residential damage was not as dramatic as in this older home.
Photo by Scott Rose

RIGHT: Rumors flew throughout the county that Sacred Heart Church was irreparably damaged. The church survived. The major problem encountered was this shift in the bell tower.
Photo by Paul Carbone

ABOVE and RIGHT: The Hollister City Council, meeting in emergency session with only one light powered by an emergency generator, was faced with deciding whether or not to let business owners back into damaged buildings. As in every other city racked by the quake, the council had to weigh the rights of business people with the concern for public safety. Tempers flared at the meetings, with merchants demanding to be let into their buildings to retrieve inventory and records. The council decided on a system to allow merchants into their stores for a specified length of time.
Photos by Don Smith, Free Lance

Tom Barry, *Hollister Building Director*

"We caught a lot of flack from some (building) owners. . . . A lot of people didn't understand what was going on. They thought we were tearing down buildings right and left, which wasn't true.

"The hard part for a person dealing with the public is all the emotion, because you hear it from everybody."

Matt Escover, *City Councilman*

"I don't think there is anything more frustrating than when people are asking you questions and you don't have the information to answer them, and the people you rely on are still trying to find out what's going on, too.

"The thing with the whole crisis was, it kept unfolding and unfolding and getting more complicated as time went on."

ABOVE: The I.O.O.F. building sits quietly, cleaned up and waiting for the bulldozer.
Photo by Rickey Popplewell

FACING PAGE: Assemblyman Rusty Areias, right, speaks with gym owner Steve Rovella, whose business was located in the severely damaged I.O.O.F. building.
Photo by Scott Rose

LEFT: Friends and employees frantically disassemble equipment in Rovella's gym prior to the building's demolition. Much of the weight-lifting machinery was hand-built.
Photo by Ray Pierce, Free Lance

Steve Rovella, *owner of Rovella's Gym, formerly in the I.O.O.F. building*

"I don't care so much about losing my business or the building, but I was surprised and happy that there was no human life lost."

ABOVE: Demolition of the venerable I.O.O.F. building on Fourth Street was swift and complete.
Photo by Don Smith, Free Lance

LEFT: The falling brick facade of the I.O.O.F. building crushed this pickup parked on Fourth Street.
Photo by Julia Lenda

BELOW: Across the street, a crowd gathered to watch the I.O.O.F. building come down.
Photo by Free Lance

ABOVE: Cleanup at the I.O.O.F site was not as dramatic as the building's demolition.
Photo by Victor Cauhape

Patty Cracolice, *owner of Baroness Beauty Supply, formerly at 525 San Benito Street*

"The most dramatic part was when they tore the building down. It's kind of difficult to watch four and a half years being destroyed in an hour, or whatever it took.

"Plus, in taking out the front half of the store, they took all my records. I had my filing cabinets there with all my invoices. I ran my business out of this cart with my filing cabinet.

"Even the shelf that was left standing . . . and two oak makeup cases I had made for the store, those were in such bad shape I couldn't salvage them. If it wasn't from the backhoe, they had water damage. You just couldn't salvage them."

ABOVE: The Baroness Beauty Supply building was one of two torn down by the City of Hollister under the directive that it posed a "clear and present danger."
Photo by Free Lance

ABOVE: Walls never meant to see the light of day are exposed as the building housing Baroness Beauty Supply is demolished for reasons of public safety.
Photo by Michael Neill

BELOW: A bulldozer completes what the earthquake began.
Photo by Free Lance

FACING PAGE: Water was suddenly worth standing in line for. Some areas of Hollister went without city water for nearly a week.
Photo by Jim McCann, San Benito Sun

BELOW LEFT: The lines moved slowly, but everyone eventually went home with containers full of water.
Photo by Dennis Taylor, Free Lance

BELOW RIGHT: Cans of drinking water were trucked in from the Anheuser Busch plant in Fairfield.
Photo by Jim McCann

ABOVE: Young L.A. Raiders football fans fill their bottles.
Photo by Ray Pierce, Free Lance

Help arrives

BELOW: There was a sense of urgency as well as community spirit as volunteers pitched in at the Red Cross Shelter established at San Andreas School.
Photo by Jim McCann, San Benito Sun

LEFT: San Andreas School served as an emergency shelter after the Veterans Memorial Building, San Benito High School and Sacred Heart School gyms were ruled out due to possible structural damage.
Photo by Jim McCann, San Benito Sun

RIGHT: Young volunteers from the League of United Latin American Citizens youth group helped sort and prepare food.
Photo by Mickie Luna

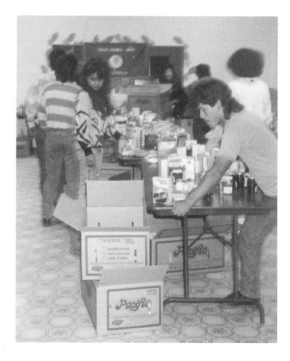

When help arrived, it came in droves. By the time major relief efforts were tapering off, close to 3,000 volunteer hours were provided by local service organizations alone. From Rotarians to Lions, Kiwaniannes to Masons, service clubs contributed nearly $300,000 to fund the recovery process of local quake victims.

LEFT: The Salvation Army was one of the many organizations which responded quickly to the needs of earthquake victims.
Photo by Eric Raper

BELOW: Residents line up at a Salvation Army assistance vehicle parked at the Hollister Community Center.
Photo by Ray Pierce, Free Lance

Ray Sanchez, *Hollister, was a volunteer during the earthquake, and is now employed by the Red Cross Tri-County Agency in Watsonville*

"Nobody had a great deal of experience with a 7.1 [earthquake] before, and so it was a disaster for all of us. I've always said that first we had the earthquake and then the disaster, because we weren't prepared as a community. It certainly has been a learning situation for all of us."

ABOVE: Ray Sanchez was one of the Red Cross volunteers who stepped in to help organize the local relief effort following the earthquake.
Photos by Free Lance

LEFT: Emergency supplies poured in to disaster stricken areas from all over the country.
Photo by Free Lance

BELOW: The Red Cross service center, originally set up as a shelter at San Andreas High School, moved to Sunnyslope Christian Center the first weekend following the quake.
Photo by Free Lance

FACING PAGE, TOP:
Volunteers were needed to load and unload the trucks full of donated items.
Photo by Jim McCann,
San Benito Sun

FACING PAGE, BOTTOM:
Cases of milk and other perishables had to be distributed quickly.
Photo by Free Lance

Kuni Wubbels, *a volunteer with other Lioness Club members at the Sunnyslope Christian Center Red Cross service center*
"They had a chef Sunday night for cooking dinner and so what they were doing was asking if somebody could make breakfast in the morning. I said, 'Well, I can make an egg for my kids, I might as well fry breakfast for whoever is coming here.' And then there was another lady by the name of Kimberly, and she was a volunteer who came from Sacramento to help out here. The two of us for a week long did the cooking."

LEFT: Richard Souza and Kathy Carnes drove from Grants Pass, Oregon with a load of much-needed plywood. The wood was donated by Souza's employer, Timber Products Co. in Central Point, Oregon. Hollister Public Works Director Jack LaPorte immediately used the material to construct needed barricades and to shore up damaged buildings.
Photo by Craig Cochran, San Benito Sun

BELOW: Donated clothing arrived by the truckload. Sunnyslope Christian Center became a temporary sorting area.
Photo by Free Lance

BELOW: Salvation Army volunteers sort through
donated clothing at the distribution center set up at
Iglesia de Dios on Nash Road.
Photo from Ken Claeys, Salvation Army

Dolores Lapp, *Salvation Army Disaster Director, Santa Clara, Santa Cruz, San Benito Counties*

"At the time of the disaster, everybody loses all the common sense they have. Everything goes out the window. You learn all this stuff and everything goes out the window. But I think it was handled fairly well and I think everybody learned from it, and that's important."

ABOVE: The California Department of Forestry and Fire Protection's emergency mess area for personnel and volunteers. CDF units from as far away as Parkfield, San Luis Obispo, and Bradley responded to the Gabilan Unit's calls. Two strike teams were formed, each consisting of five engines with a team of three and a unit leader in a separate vehicle.
Photo by Ray Pierce, Free Lance

Dan Holsapple, *Hollister Fire Chief*

"The Department of Forestry's strike team arrived the second day and we brought down an engine into the city to assist our guys downtown. We cut the town in half, so everything east of San Benito Street was covered by the Department of Forestry, and everything to the west side of town and into the industrial park, the airport, and all that, the high school, that area was covered by the city units."

RIGHT: Ron Ross, Hollister Fire Department volunteer, grabs a quick bite at the CDF staging area. *Photo by Ray Pierce, Free Lance*

ABOVE: A Salinas Hazardous Materials squad was one of a number of regional fire department teams which lent support to local emergency repair and damage assessment efforts. *Photo by Dennis Taylor, Free Lance*

BELOW: PG & E crews from Utah, Nevada, southern California and other communities came to assist local personnel with the monumental tasks facing them following the quake. *Photo by Free Lance*

ABOVE: San Benito County was assisted by law enforcement officers from nearby jurisdictions, including units from Monterey, Carmel, Salinas, Gilroy, Morgan Hill and the California Highway Patrol.
Photo by Eric Raper

Hugh Riley, *Hollister City Manager*
"One of the remarkable things that occurred, basically almost instantly, was the law enforcement assistance we received from the outside. It was a big help."

John Wrobel, *amateur radio operator, on emergency communication problems and how local "Hams" assisted disaster officials*
"The Fremont Peak repeater was down, so all [radio operators] could only talk locally.

"After the quake, I went up to the communications center, and I met up with Harry and Kathy Hill. Harry set up a radio in a car (using the battery for power) and started talking to people. I went back home to get a power supply.

"We sent Kathy over to the hospital, several people we sent over to the shelter. We provided information to the state Office of Emergency Services about how many people were in the shelters, at the hospital. At one point we sent a guy over to Mission Oaks Mobile Home Park.

"Harry and Kathy were real troupers, they were up there all the time."

LEFT: Exhausted California Conservation Corps workers take a much needed break on the lawn at Dunne Park.
Photo by Dennis Taylor, Free Lance

LEFT: Hollister Airport was busy with helicopters and other transports bringing supplies, donations and VIP's to San Benito County.
Photo by Free Lance

LEFT: California Governor George Deukmejian, center, toured downtown Hollister twice in the months following the earthquake. On his second trip, in December, the Governor announced the possibility for the state to purchase quake-damaged product from Hollister canneries. Thousands of pounds of canned tomato goods were dented but not broken, but could not be sold because they weren't labeled prior to the quake. The plan originated in the office of Assemblyman Rusty Areias, left. Greg Light, right, appointed mayor in November, also accompanied the Governor on his tour. Areias was later appointed the chairman of the Assembly Earthquake Preparedness Committee.
Photo by Ray Pierce, San Benito Sun

David Edge, *County Administrative Officer and Director of Emergency Services, on Governor Deukmejian's visits*
"He was clearly personally moved by what he'd seen, and was very responsive to the points we were putting across."

RIGHT: On Governor Deukmejian's first visit to Hollister, in the days immediately following the earthquake, he saw the effects of the temblor first hand. Dean Hallberg, left, was Hollister's mayor at the time of the quake, and Mike Graves, right, was the chairman of the County Board of Supervisors.
Photo by Free Lance

ABOVE: Marilyn Quayle, wife of Vice-President Dan Quayle, swung through Hollister on a fact-finding tour of earthquake damaged areas. A disaster relief specialist for the Bush administration, Quayle was accompanied in Hollister by, from left to right, Hollister Building Director Tom Barry, Hollister Downtown Association President Bob Tiffany, Hollister City Councilman Matt Escover, and at right, Sylvia Panetta, wife and aide to Congressman Leon Panetta.
Photo by Jim McCann, San Benito Sun

David Edge, *San Benito County Administrative Officer and Director of Emergency Services*

"In our case, if it hadn't been for [U.S. Rep.] Leon Panetta and his staff, notably his wife, and their untiring efforts to help us through the Federal Bureacracy, there would have been some real recovery problems, especially with getting emergency shelters for people who lost their homes."

ABOVE: County Administrative Officer David Edge briefs U.S. Representative Leon Panetta on the status of recovery operations. Panetta hosted a series of Saturday meetings and acted as an intermediary between local officials and federal disaster aid personnel.
Photo by Jim McCann, San Benito Sun

BELOW: Hollister's Mayor at the time of the earthquake, Dean Hallberg, found himself a regular on the nightly news.
Photo by Jim McCann, San Benito Sun

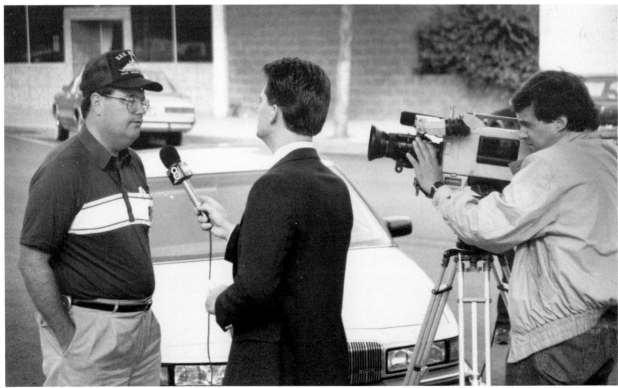

EMERGENCY
SERVICE

* County Tax Relief
* Mental Health
* CA. Employment Dept.
* Salvation Army
* IRS

CENTER
MANAGER

GERENTE
DEL
CENTRO

ABOVE: Nights, for many people, were the hardest times. The memories of the quake, of houses rocking, windows breaking, the loss of water and power, were strongest after dark.
Photo by Ray Pierce, Free Lance

FACING PAGE, TOP: The Hollister Community Center became the home of the local Disaster Application Center.
Photo from Ken Claeys

FACING PAGE, BOTTOM: Those needing financial aid came face to face with bureaucracy and the proverbial red tape.
Photos by Robert Ojeda

Dennis Taylor, *reporter for the Free Lance newspaper*

"Short of covering plane crashes or wars, no one in this area had ever covered anything like that earthquake before. It was new ground. I didn't know how I was going to cover it until I got out there.

"Basically what I found was, there is no way to cover it. 'It' itself is the story and you just have to make yourself available to assimilate all the information coming in from the different perspectives: the human tragedy, the community tragedy, the loss of homes, loss of businesses, the pain, the fear; the look of fear you see in children's faces sitting on a curb at night, when their parents are trying to board up their business.

"It's just a matter of being part of the same community that was hit, and reporting it. You're not feeling any different than anyone else. I felt no less fear than the child clutching a doll sitting on a sidewalk."

Les Trevino, *plant manager at Shelton Industries, recalls the moments following the quake. The roof had collapsed over the motor coach in which he and other employees were working*

"We were all in shock, wondering what was the next thing to do. Within seconds everyone started dashing. Fortunately the doors were wide open still. When we looked back, we could still hear the rumbling, see the dust settling."

Industry Hit Hard

BELOW: The weight of thousands of cans caused extensive damage to area storage facilities. Both of Hollister's canneries suffered quake-caused losses.
Photo by Ray Pierce, Free Lance

ABOVE and LEFT: Workers at Shelton Industries, including Les Trevino, were inside this trailer when the cans stacked next door pushed through the adjoining wall and pulled the ceiling down.
Photos by Ray Pierce, Free Lance

RIGHT: The overwhelming job of sorting through fallen stacks of canned goods had to be completed one can at a time.
Photo by Danny Valencia

BELOW: The north wall of this cannery warehouse was ripped away during the quake.
Photo by Rickey Popplewell

ABOVE: Sorting through mounds of cans, Tri-Valley
Growers workers found a crushed car buried underneath.
No one was inside the vehicle.
Photo by Jim McCann, San Benito Sun

RIGHT: Close-up of another
vehicle crushed by an avalanche
of canned goods.
Photo by Danica Johnson

1906 Remembered

It happened before. The great earthquake of April 18, 1906 devastated downtown Hollister. Ironically, these photos, taken in the aftermath of that historic event, lay hidden until the Loma Prieta quake. They were discovered in Margaret Houghton's historic Hawkins house on South Street while cleaning up the debris thrown on October 17, 1989.

Hollister resident Bernice Palmtag Day remembers the morning in April 1906 when the earth trembled. Born at the Cienega Winery, Mrs. Day was a young girl living with her family in a house there. The chimney collapsed, as did most in Hollister, and the bricks fell through the roof onto her bed. The weight broke the casters. Her mother grabbed her three-month-old baby sister, Louise Palmtag Mesquit, putting her head on her left shoulder when a falling beam struck her on the right shoulder - sparing her life and her baby's. Cut by the falling bricks, Bernice needed stitches. Father Palmtag drove her 10 miles to town in a buggy, followed by a man on horse back to swim her across the San Benito River if it had washed out the bridge.

Clearing the rubble in 1906, as in the photograph above, was accomplished with the aid of a horse and buggy.

A building's method of construction has much to do with its ability to withstand an earthquake. It was true in 1906, as the photo at right attests, and it holds true today.

Photos courtesy of Margaret Houghton

An orphanage operated by an order of nuns, the grand house at right succumbed to the 1906 earthquake. The skewed angles of once-right walls is reminiscent of the damage many older homes received in October 1989.

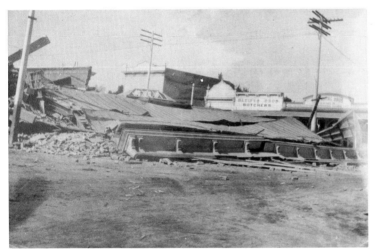

The great quake literally flattened buildings in downtown Hollister. Not all was disaster, however. Out in Cienega, Bernice Palmtag Day recalls the wine draining from damaged barrels, running down the street through the pig pens. "The pigs were drunk for days," she said.

But for the style of clothes, the old-fashioned baby stroller and the lack of asphalt, the scene at right is almost identical to San Benito Street immediately following the Loma Prieta quake. Residents were stunned and, at the same time, curious.

History is said to repeat itself. Earthquakes in San Benito County, and elsewhere along the faults, seem to prove the point.

The Earthquake Capital of the World

By Jeff Tolhurst

"Civilization exists by geological consent subject to change without notice."
-Will Durant

History and Geology

The geology of San Benito County plays a very important role in the lives of the county's residents. Whether it be minerals or earthquakes the county is unique in California. The area's natural resources have drawn, and continue to draw, many people here. Mercury ore was first discovered in San Benito County at New Idria, 67 miles southeast of Hollister, bringing some of the first settlers in 1853. The quicksilver (mercury) was used by the prospectors to separate gold found in the Sierras from less valuable rock material. Other mining activities continue in San Benito County, including sand and gravel, gems such as Benitoite, dolomite, oil, and gas to name a few. The location of geologic resources has helped shape San Benito County and continues to do so today.

Plate Tectonics

The geology of San Benito County can be explained by a relatively new theory called plate tectonics. The main idea behind the theory states that the outer layer of the earth is composed of large, rigid rock slabs call plates, which can be up to 50 miles in thickness. These plates lie on another layer that is softer and can flow to a certain degree (similar to toothpaste).

Heat sources from within the earth cause molten rock material to rise toward the surface. The currents that form are known as convection currents and are the forces that drive the rigid plates across the earth. As the plates move, they rub against one another, causing earthquakes along their boundaries. Most of the world's seismic and volcanic activity occurs along plate boundaries.

In general, there are three types of plate boundaries: diverging, converging, and transform. Diverging plate boundaries occur as magma rises up between two plates, forcing them apart. Converging plate boundaries occur as two plates are forced together. Transform plate boundaries occur as two plates grind past one another.

As new earth material forms along divergent plate boundaries, the earth's crust is forced apart. This is happening along the mid-oceanic ridges as lava cools into new rock. It is also occurring off the coast of Northern California, and between Baja California and mainland Mexico - which has implications for California, located in between.

Mountain ranges, such as the Andes and the Himalayas, are typically found along convergent plate boundaries. The forces causing plates to spread apart also causes the opposite sides of the plates to smash into one another. If two continental plates collide, mountains form. If two oceanic plates collide, a process termed subduction typically occurs, where one plate is forced under the other. Subduction also occurs when an oceanic plate is forced under a continental plate. Oceanic plate material is typically more dense than continental plate material and has a tendency to sink along convergent plate boundaries.

The San Andreas fault system is an example of a transform plate boundary. The two plates forming the boundary are slipping past one another as they move in different directions. This movement can be very subtle or it can be very violent, as evidenced by the Loma Prieta earthquake.

San Andreas Fault System

San Benito County is famous for its earthquakes. These occur because part of the San Andreas fault system runs lengthwise through the county. It demarks the border between the Pacific and North American plates. The rate at which the plates move relative to one another averages a few centimeters a year. Over millions of years, however, this distance adds up. Some geologists believe the Gabilan mountain range has moved from southeastern California to its present position south and west of Hollister and San Juan Bautista over the last 200 million years.

The fault was studied as early as 1891 and the name was apparently first used by A.C. Lawson in 1895. After the 1906 San Francisco earthquake, scientists recognized the magnitude and extent of the fault. Reports described a displacement of almost a foot along the fault just to the northwest of San Juan Bautista.

Some geologists have divided the fault into segments according to the amount of seismic activity along its length (See figure 1). The segment through San Benito County is said to be "creeping" or moving along at a steady rate rather than being "locked" and moving only occasionally in large amounts (as in 1906 and 1989). This was first noticed on the San Andreas at the

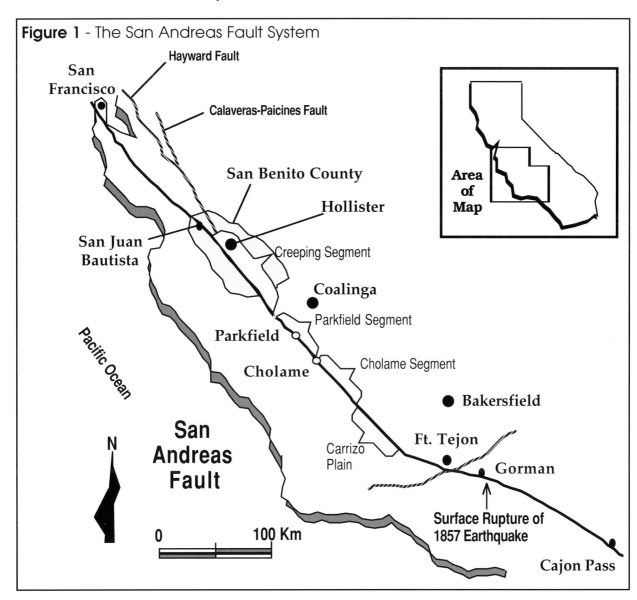

Figure 1 - The San Andreas Fault System

Hayward Fault

San Francisco

Calaveras-Paicines Fault

San Benito County

Hollister

San Juan Bautista

Creeping Segment

Coalinga

Parkfield Segment

Parkfield

Cholame Segment

Cholame

Bakersfield

Pacific Ocean

San Andreas Fault

N

Carrizo Plain

Ft. Tejon

Gorman

Surface Rupture of 1857 Earthquake

Cajon Pass

0   100 Km

Area of Map

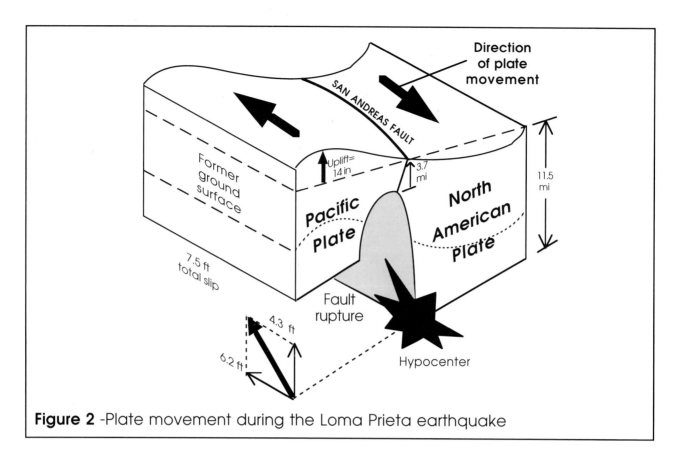

Figure 2 -Plate movement during the Loma Prieta earthquake

Almaden Winery on Cienega Road, south of Hollister, then on the Calaveras fault in Hollister.

The Calaveras fault trace runs from Nash Road just west of San Benito High School through Hollister past the west side of Park Hill. The Calaveras fault is presumed to connect with the San Andreas fault south of Hollister. As it proceeds northward, it appears to link with the Hayward fault north of Morgan Hill. The Sargent fault appears to connect the San Andreas fault with the Calaveras fault north of Hollister and San Juan Bautista and south of Gilroy. San Benito County is cut by numerous other faults whose presence indicate past seismic activity within the region. Because Hollister has so many earthquakes annually, it has become known to some as the Earthquake "Capital of the World." There are other places that have more seismic events per year, but Hollister is one of the few places in the world where there is actual detectable movement along a fault. Some geologists believe the release of energy

by these small to medium magnitude events may help to decrease the chances of a very large magnitude earthquake having an epicenter in, or very near, Hollister. On the other hand, stress seems to be building up to the north along the San Andreas fault as the Pacific plate is forced to the northwest. The United States Geological Survey says there is a 67 percent chance of another earthquake the size of Loma Prieta during the next 30 years somewhere between San Jose and Santa Rosa on either side of San Francisco Bay. It could strike at any time. Although the epicenter may be over 50 miles from San Benito County, residents could experience some very serious shaking. We may have active creep and many smaller quakes, but we are not immune to the effects of ground motion from a larger, more distant earthquake.

Loma Prieta Earthquake

The Loma Prieta earthquake of October 17, 1989 was the largest seismic event to occur along the segment of the San Andreas fault which extends from San